Contents

Introduction

1　This booklet gives general guidance on the Manual Handling Operations Regulations 1992 (the Regulations) which came into force on 1 January 1993. The Regulations are made under the Health and Safety at Work etc Act 1974 (the HSW Act). They implement European Directive 90/269/EEC on the manual handling of loads (see reference section at the back of this publication), supplement the general duties placed upon employers and others by the HSW Act and the broad requirements of the Management of Health and Safety at Work Regulations 1992 (see reference section), and replace a number of earlier, outdated legal provisions.

2　More than a quarter of the accidents reported each year to the enforcing authorities are associated with manual handling - the transporting or supporting of loads by hand or by bodily force. While fatal manual handling accidents are rare, accidents resulting in a major injury such as a fractured arm are more common, accounting for 6% of all major injuries reported in 1990/91. The vast majority of reported manual handling accidents result in over-three-day injury, most commonly a sprain or strain, often of the back. Figures 1 to 3 illustrate these patterns for over-three-day injuries reported in 1990/91.

3　Sprains and strains arise from the incorrect application and/or prolongation of bodily force. Poor posture and excessive repetition of movement can be important factors in their onset. Many manual handling injuries are cumulative rather than being truly attributable to any single handling incident. A full recovery is not always made; the result can be physical impairment or even permanent disability.

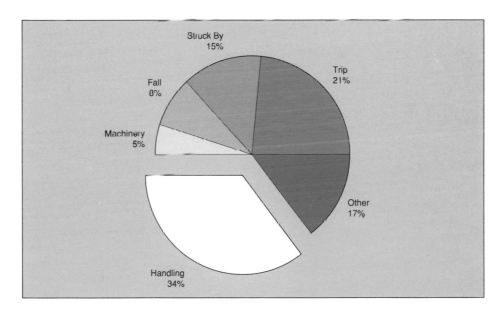

Fig 1: Kinds of accident causing injury

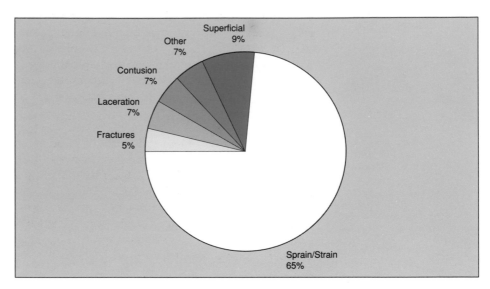

Fig 2: Types of injury caused by handling accidents

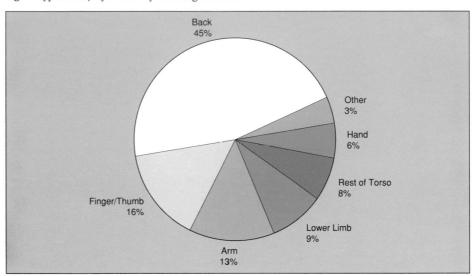

Fig 3: Sites of injuries caused by handling

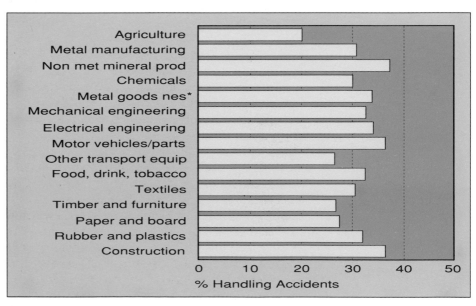

*nes: not elsewhere specified

Fig 4: Percentage of injuries caused by handling

4   Figure 4, also based on over-three-day injuries reported in 1990/91, shows that the problem of manual handling is not confined to a narrow range of industries but is widespread. Nor should it be supposed that the problem is confined to 'industrial' work: for example the comparable figure for both banking and finance and for retail distribution is 31% and for medical, veterinary and other health services 55%.

5   There is now substantial international acceptance of both the scale of the manual handling problem and methods of prevention. Modern medical and scientific knowledge stresses the importance of an ergonomic approach in removing or reducing the risk of manual handling injury. Ergonomics is sometimes described as 'fitting the job to the person, rather than the person to the job'. The ergonomic approach therefore looks at manual handling as a whole, taking into account a range of relevant factors including the nature of the task, the load, the working environment and individual capability. This approach is central to the European Directive on manual handling, and to the Regulations.

6   The Regulations should not be considered in isolation. Regulation 3(1) of the Management of Health and Safety at Work Regulations 1992 requires employers to make a suitable and sufficient assessment of the risks to the health and safety of their employees while at work. Where this general assessment indicates the possibility of risks to employees from the manual handling of loads the requirements of the present Regulations should be followed.

7   The Regulations establish a clear hierarchy of measures:

(a)   avoid hazardous manual handling operations so far as is reasonably practicable - this may be done by redesigning the task to avoid moving the load or by automating or mechanising the process;

(b)   make a suitable and sufficient assessment of any hazardous manual handling operations that cannot be avoided; and

(c)   reduce the risk of injury from those operations so far as is reasonably practicable - particular consideration should be given to the provision of mechanical assistance but where this is not reasonably practicable then other improvements to the task, the load and the working environment should be explored.

8   Like the European Directive on manual handling, the Regulations set no specific requirements such as weight limits. The ergonomic approach shows clearly that such requirements are based on too simple a view of the problem and are likely to lead to erroneous conclusions. Instead, an ergonomic assessment based on a range of relevant factors is used to determine the risk of injury and point the way to remedial action.

9   However, a full assessment of every manual handling operation could be a major undertaking and might involve wasted effort. Therefore Appendix 1 offers numerical guidelines which can be used as an initial filter, helping to identify those manual handling operations which warrant a more detailed examination. The guidelines set out an approximate boundary within which manual handling operations are unlikely to create a risk of injury sufficient to warrant more detailed assessment. This should enable assessment work to be concentrated where it is most needed. However even operations lying within the boundary should be avoided or made less demanding wherever it is reasonably practicable to do so. **The guidelines should not be regarded as precise recommendations. They should be applied with caution. Where doubt remains a more detailed assessment should be made.**

10   It is intended that the contents of this booklet should form a general framework within which individual industries and sectors will be able to produce more specific guidance appropriate to their own circumstances.

11   Manual handling injuries are part of a wider family of musculoskeletal problems; the reader may also find it helpful to refer to the Health and Safety Executive's (HSE) booklet *Work related upper limb disorders - a guide to prevention* (see reference section at the back of this publication).

FLOW CHART

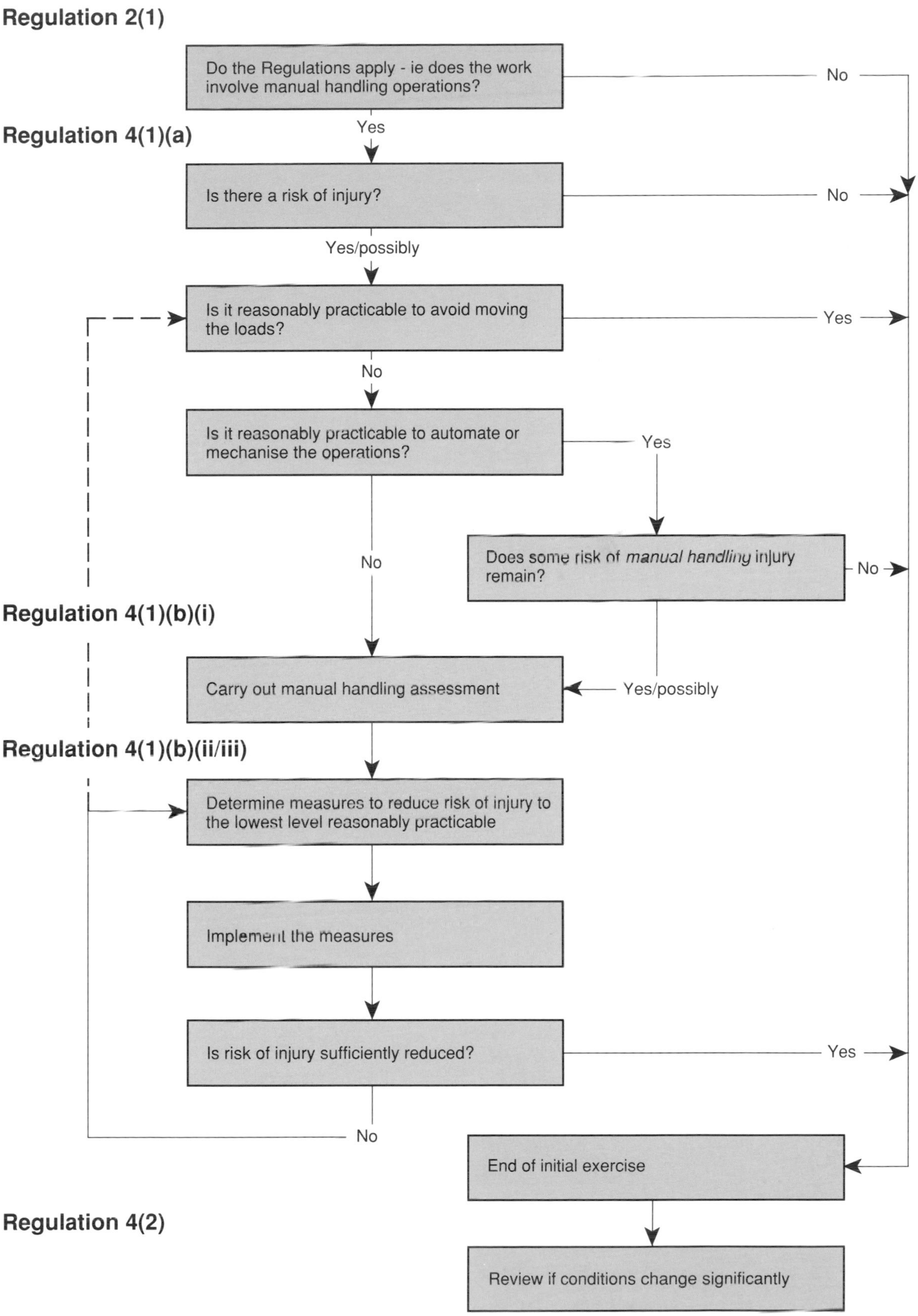

Regulation 2(1)

Do the Regulations apply - ie does the work involve manual handling operations? → No

Regulation 4(1)(a)

↓ Yes

Is there a risk of injury? → No

↓ Yes/possibly

Is it reasonably practicable to avoid moving the loads? → Yes

↓ No

Is it reasonably practicable to automate or mechanise the operations? → Yes

↓ Yes

Does some risk of *manual handling* injury remain? → No

↓ No

Regulation 4(1)(b)(i)

Carry out manual handling assessment ← Yes/possibly

Regulation 4(1)(b)(ii/iii)

↓

Determine measures to reduce risk of injury to the lowest level reasonably practicable

↓

Implement the measures

↓

Is risk of injury sufficiently reduced? → Yes

↓ No

End of initial exercise

Regulation 4(2)

↓

Review if conditions change significantly

Fig 5: How to follow the Manual Handling Operations Regulations 1992

Citation and commencement

These Regulations may be cited as the Manual Handling Operations Regulations 1992 and shall come into force on 1st January 1993.

Regulation 2

Interpretation

- (1) In these Regulations, unless the context otherwise requires -

"injury" does not include injury caused by any toxic or corrosive substance which -

(a)   has leaked or spilled from a load;

(b)   is present on the surface of a load but has not leaked or spilled from it; or

(c)   is a constituent part of a load;

and "injured" shall be construed accordingly;

"load" includes any person and any animal;

"manual handling operations" means any transporting or supporting of a load (including the lifting, putting down, pushing, pulling, carrying or moving thereof) by hand or by bodily force.

Definitions of certain terms

Injury

12   The Regulations seek to prevent injury not only to the back but to any part of the body. Account should be taken of any external physical properties of loads which might either affect grip or cause direct injury, for example slipperiness, roughness, sharp edges, extremes of temperature.

13   Hazards from toxic or corrosive properties of loads through spillage or leakage or from external contamination are not covered by these Regulations, though such hazards should be considered in the light of other provisions such as COSHH - the Control of Substances Hazardous to Health Regulations 1988 (see reference section). For example, the presence of oil on the surface of a load is relevant to the Regulations if it makes the load slippery to handle; but a risk of dermatitis from contact with the oil is dealt with elsewhere.

Load

14   A load in this context must be a discrete moveable object. This includes, for example, a human patient receiving medical attention or an animal during husbandry or undergoing veterinary treatment, and material supported on a shovel or fork. An implement, tool or machine - such as a chainsaw - is not considered to constitute a load while in use for its intended purpose.

Manual handling operations

15   The Regulations apply to the manual handling of loads, ie by human effort, as opposed to mechanical handling by crane, lift truck, etc. The human effort may be applied directly to the load, or indirectly by hauling on a rope or pulling on a lever. Introducing mechanical assistance, for example a sack truck or a powered hoist, may reduce but not eliminate manual handling since human effort is still required to move, steady or position the load.

16   Manual handling includes both transporting a load and supporting a load in a static posture. The load may be moved or supported by the hands or any other part of the body, for example the shoulder. Manual handling also includes the intentional dropping of a load and the throwing of a load, whether into a receptacle or from one person to another.

17   The application of human effort for a purpose other than transporting or supporting a load does not constitute a manual handling operation. For example turning the starting handle of an engine or lifting a control lever on a machine is not manual handling; nor is the action of pulling on a rope while lashing down cargo on the back of a vehicle.

Regulation
2(2)

(2)   Any duty imposed by these Regulations on an employer in respect of his employees shall also be imposed on a self-employed person in respect of himself.

Duties of the self-employed

18   Regulation 2(2) makes the self-employed responsible for their own safety during manual handling. They should take the same steps to safeguard themselves as would be expected of employers in protecting their employees in similar circumstances.

Regulation 3

Disapplication of Regulations

Regulation
3

These Regulations shall not apply to or in relation to the master or crew of a sea-going ship or to the employer of such persons in respect of the normal ship-board activities of a ship's crew under the direction of the master.

Sea-going ships

19   Sea-going ships are subject to separate Merchant Shipping legislation administered by the Department of Transport. The Regulations therefore do not apply to the normal ship-board activities of a ship's crew under the direction of the master. However the Regulations may apply to other manual handling operations aboard a ship, for example where a shore-based contractor carries out the work, provided the ship is within territorial waters. The Regulations also apply to certain activities carried out offshore - see regulation 7.

Regulation 4

Duties of employers

Introduction

20   The present Regulations should not be considered in isolation. Regulation 3(1) of the Management of Health and Safety at Work Regulations 1992 (see reference section at the back of this publication) requires employers to make a suitable and sufficient assessment of the risks to the health and safety of their employees while at work. Where this general assessment indicates the possibility of risks to employees from the manual handling of loads the requirements of the present Regulations should be observed, as follows.

Hierarchy of measures

21   Regulation 4(1) establishes a clear hierarchy of measures:

(a)   avoid hazardous manual handling operations so far as is reasonably practicable;

(b)   assess any hazardous manual handling operations that cannot be avoided; and

(c)   reduce the risk of injury so far as is reasonably practicable.

Extent of the employer's duties

22   The extent of the employer's duty to avoid manual handling or to reduce the risk of injury is determined by reference to what is 'reasonably practicable'. Such duties are satisfied if the employer can show that the cost of any further preventive steps would be grossly disproportionate to the further benefit that would accrue from their introduction.

23   This approach is fully applicable to the work of the emergency services. Ultimately, the cost of prohibiting all potentially hazardous manual handling operations would be an inability to provide the general public with an adequate rescue service. A fire authority, for example, may therefore consider that it has discharged this duty when it can show that any further preventive steps would make unduly difficult the efficient discharge of its emergency functions.

A continuing duty

24   It is not sufficient simply to make changes and then hope that the problem has been dealt with. Steps taken to avoid manual handling or reduce the risk of injury should be monitored to check that they are having the desired effect in practice. If they are not, alternative steps should be sought.

25   It should also be remembered that regulation 4(2) (discussed later) requires the assessment made under regulation 4(1) to be kept up to date.

Work away from the employer's premises

26   The Regulations impose duties upon the employer whose employees carry out the manual handling. However, manual handling operations may occur away from the employer's premises in situations over which little direct control can be exercised. Where possible the employer should seek close liaison with those in control of such premises. There will sometimes be a limit to employers' ability to influence the working environment; but the task and perhaps the load will often remain within their control, as will the provision of effective training, so it is still possible to establish a safe system of work.

27   Employers and others in control of premises at which visiting employees have to work also have duties towards those employees, particularly under sections 3 or 4 of the HSW Act, the Management of Health and Safety at Work Regulations 1992 and the Workplace (Health, Safety and Welfare) Regulations 1992 (see reference section), for example to ensure that the premises and plant provided there are in a safe condition.

- (1) Each employer shall -

(a)   so far as is reasonably practicable, avoid the need for his employees to undertake any manual handling operations at work which involve a risk of their being injured.

Avoidance of manual handling

Risk of injury

28   If the general assessment carried out under regulation 3(1) of the Management of Health and Safety at Work Regulations 1992 (see reference section at the back of this publication) indicates a possibility of injury from manual handling operations, consideration should first be given to avoiding the need for the operations in question. At this preliminary stage a judgement should be made as to the nature and likelihood of injury. It may not be necessary to assess in great detail, particularly if the operations can readily be avoided or if the risk is clearly of a low order. Appendix 1 provides some simple numerical guidelines to assist with this initial judgement, at least in relatively straightforward cases.

Elimination of handling

29   In seeking to avoid manual handling the first question to ask is whether movement of the loads can be eliminated altogether: are the handling operations unnecessary; or could the desired result be achieved in some entirely different way? For example, can a process such as machining or wrapping be carried out in situ, without handling the loads? Can a treatment be brought to a patient rather than taking the patient to the treatment?

Automation or mechanisation

30   Secondly, if load handling operations, in some form, cannot be avoided entirely then further questions should be asked:

(a)   can the operations be automated?

(b)   can the operations be mechanised?

31   It should be remembered that the introduction of automation or mechanisation may create other, different risks. Even an automated plant will require maintenance and repair. Mechanisation, for example by the introduction of lift trucks or powered conveyors, can introduce fresh risks requiring precautions of their own.

32   It is especially important to address these questions when plant or systems of work are being designed. However, examination of existing activities may also reveal opportunities for avoidance of manual handling operations that involve a risk of injury. Such improvements often bring additional benefits in terms of greater efficiency and productivity, and reduced damage to loads.

- (1) Each employer shall -

    (b)   where it is not reasonably practicable to avoid the need for his employees to undertake any manual handling operations at work which involve a risk of their being injured -

        (i)   make a suitable and sufficient assessment of all such manual handling operations to be undertaken by them, having regard to the factors which are specified in column 1 of Schedule 1 to these Regulations and considering the questions which are specified in the corresponding entry in column 2 of that Schedule.

Assessment of risk

33   Where the general assessment carried out under regulation 3(1) of the

Management of Health and Safety at Work Regulations 1992 (see reference section) indicates a possibility of injury from manual handling operations but the conclusion reached under regulation 4(1)(a) is that avoidance of the operations is not reasonably practicable; a more specific assessment should be carried out as required by regulation 4(1)(b)(i). The extent to which this further assessment need be pursued will depend on the circumstances. Appendix 1 offers some simple numerical guidelines to assist with this decision. The guidelines are intended to be used as an initial filter, to help to identify those operations deserving more detailed assessment.

34   Schedule 1 to the Regulations specifies factors which this assessment should take into account including the **task**, the **load**, the **working environment** and **individual capability**. First, however, consideration should be given to how the assessment is to be carried out - and by whom - and what other relevant information may be available to help.

Who should carry out the assessment?

35   In most cases employers should be able to carry out the assessment themselves or delegate it to others in their organisation, having regard to regulation 6 of the Management of Health and Safety at Work Regulations 1992 (see reference section). A meaningful assessment can only be based on a thorough practical understanding of the type of manual handling tasks to be performed, the loads to be handled and the working environment in which the tasks will be carried out. Employers and managers should be better placed to know about the manual handling taking place in their own organisation than someone from outside.

36   While one individual may be able to carry out a perfectly satisfactory assessment, at least in relatively straightforward cases, it can be helpful to draw on the knowledge and expertise of others. In some organisations this has been done informally; others have preferred to set up a small assessment team.

37   Areas of knowledge and expertise likely to be relevant to the successful assessment of risks from manual handling operations, and individuals who may be able to make a useful contribution, include:

(a)   the requirements of the Regulations (manager, safety professional);

(b)   the nature of the handling operations (supervisor, industrial engineer);

(c)   a basic understanding of human capabilities (occupational health nurse, safety professional);

(d)   identification of high risk activities (manager, supervisor, occupational health nurse, safety professional); and

(e)   practical steps to reduce risk (manager, supervisor, industrial engineer, safety professional).

38   It may be appropriate to seek outside assistance, for example to give basic training to in-house assessors or where manual handling risks are novel or particularly difficult to assess. Possible sources of such assistance are given in the reference section at the back of this document. Outside specialist advice may also help solve unusual handling problems or contribute to ergonomic design. But employers will still wish to oversee the assessment as they have the final responsibility for it.

Employees' contribution

39   The views of staff can be of particular value in identifying manual handling problems and practical solutions to them. Employees, their safety representatives and safety committees should be encouraged to play a positive part in the assessment process. They can assist the employer by highlighting difficulties arising from such things as the size or shape of loads, the frequency with which they are handled or the circumstances in which the handling operations are carried out.

Records of accidents and ill health

40   Well-kept records of accidents and ill health can play a useful part in the assessment process. They should identify accidents associated with manual handling, and careful analysis may also yield evidence of links between manual handling and ill health, including injuries apparently unrelated to any specific event or accident. Other possible indicators of manual handling problems include high levels of absenteeism or staff turnover, poor productivity and morale, excessive product damage, and general dissatisfaction among the employees concerned. Any regular occurrence of back disorders or other ailments possibly associated with unsatisfactory manual handling practices should be investigated. However such indicators are not a complete guide and should be used only to augment other risk assessment methods.

How detailed should an assessment be?

41   Employers' assessments will be 'suitable and sufficient' if they look in a considered way at the totality of the manual handling operations their employees are required to perform. Properly based 'generic' assessments which draw together common threads from a range of broadly similar operations are quite acceptable. Indeed a more narrowly focused assessment may fail to reflect adequately the range of operations encountered.

42   An assessment made at the last minute is unlikely to be 'suitable and sufficient'. In conducting assessments employers should therefore use their experience of the type of work their employees perform, consulting the employees as appropriate. This approach will help with the assessment of work that is of a varied nature (such as construction or maintenance), peripatetic (such as making deliveries) or involves dealing with emergencies (such as fire-fighting and rescue).

43   In the case of delivery operations, for example, a useful technique is to list the various types of task, load and working environment concerned and then to review a selection of them. The aim should be to establish the range of manual handling risks to which employees are exposed and then to decide on appropriate preventive steps where these are shown to be necessary.

44   A distinction should be made between the employer's assessment required by regulation 4(1)(b)(i) and the everyday judgements which supervisors and others will have to make in dealing with manual handling operations. The assessment should identify in broad terms the problems likely to arise during the kind of operations that can be foreseen and the measures that will be necessary to deal with them. These measures should include the provision of training to enable supervisors, and where appropriate individual employees, to cope effectively with the operations they are likely to undertake.

45   This distinction is perhaps most clearly seen in the case of emergency work. Here it will be essential to provide training to enable fire officers, for example, to take the rapid judgements that will inevitably be necessary in

11

dealing satisfactorily with an emergency incident or in supervising realistic training.

Industry-specific data and assessments

46   Individual industries and sectors have a valuable role to play in identifying common manual handling problems and developing practical solutions. Industry associations and similar bodies can also act as a focus for the collection and analysis of accident and ill health data drawn from a far wider base than that available to the individual employer.

Recording the assessment

47   In general, the significant findings of the assessment should be recorded and the record kept, readily accessible, as long as it remains relevant. However, the assessment need not be recorded if:

(a)   it could very easily be repeated and explained at any time because it is simple and obvious; or

(b)   the manual handling operations are quite straightforward, of low risk, are going to last only a very short time, and the time taken to record them would be disproportionate.

Making a more detailed assessment

48   When a more detailed assessment is necessary it should follow the broad structure set out in Schedule 1 to the Regulations. The Schedule lists a number of questions in five categories including the **task**; the **load**; the **working environment**; and **individual capability**. Not all of these questions will be relevant in every case.

49   These categories are clearly interrelated: each may influence the others and therefore none can be considered in isolation. However, in order to carry out an assessment in a structured way it is often helpful to begin by breaking the operations down into separate, more manageable items.

Assessment checklist

50   It may be helpful to use a checklist during assessment as an aide-memoire. An example of such a checklist is provided in Appendix 2. This checklist addresses not only the analysis of risk required by regulation 4(1)(b)(i) but also the identification of steps to reduce the risk as required by regulation 4(1)(b)(ii) discussed later. The particular example given will not be suitable in all circumstances; it can be adapted or modified as appropriate.

51   **REMEMBER** - assessment is not an end in itself, merely a structured way of analysing risks and pointing the way to practical solutions.

The task - making an assessment

Is the load held or manipulated at a distance from the trunk?

52   As the load is moved away from the trunk the general level of stress on the lower back rises. Regardless of the handling technique used, failure to keep the load close to the body will increase the stress. As a rough guide holding a load at arm's length imposes about five times the stress experienced when holding the same load very close to the trunk.

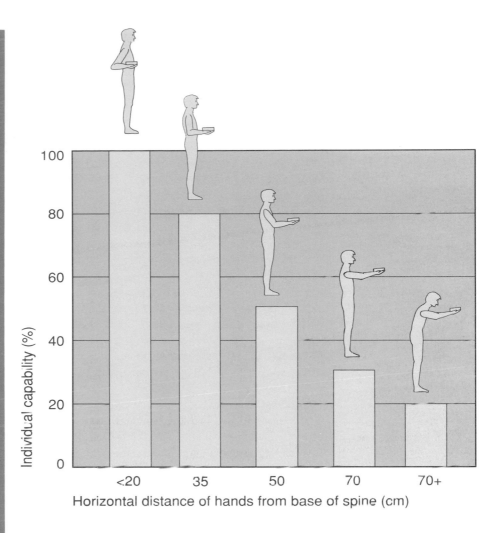

Fig 6: Reduction of individual handling capability as the hands move away from the trunk

53   In addition, the further away the load, the less easy it is to control. The benefit of friction between the load and the worker's garments in helping to support or steady the load is reduced or lost, and it is more difficult to counterbalance the load with the weight of the trunk.

The importance of posture

54   Poor posture during manual handling introduces the additional risk of loss of control of the load and a sudden, unpredictable increase in physical stresses. The risk of injury is increased if the feet and hands are not well placed to transmit forces efficiently between the floor and the load. A typical example of this is when the body weight is forward on the toes, the heels are off the ground and the feet are too close together.

Does the task involve twisting the trunk?

55   Stress on the lower back is increased significantly if twisted trunk postures are adopted. Still worse is to twist while supporting a load.

Does the task involve stooping?

56   Stooping can also increase the stress on the lower back. This happens whether the handler stoops by bending the back or by leaning forward with the back straight - in each case the trunk is thrown forward and its weight is added to the load being handled.

Does the task involve reaching upwards?

57   Reaching upwards places additional stresses on the arms and back. Control of the load becomes more difficult and, because the arms are extended, they are more prone to injury.

The effect of combining risk factors

58   Individual capability can be reduced substantially if twisting is combined with stooping or stretching. Such combinations should be avoided wherever possible, especially since their effect on individual capability can be worse than the simple addition of their individual effects might suggest. A requirement to position the load with precision can also add to the risk of injury.

Does the task involve excessive lifting or lowering distances?

59   The distance through which a load is lifted or lowered can also be important: large distances are considerably more demanding physically than small ones. Moreover lifting or lowering through a large distance is likely to necessitate a change of grip part way, further increasing the risk of injury. Lifts commencing at floor level should be avoided where possible; where unavoidable they should preferably terminate no higher than waist height.

Does the task involve excessive carrying distances?

60   In general, if a load can safely be lifted and lowered, it can also be carried without endangering the back. However, if a load is carried for an excessive distance, physical stresses are prolonged leading to fatigue and increased risk of injury. As a rough guide if a load is carried further than about 10 m then the physical demands of carrying the load will tend to predominate over those of lifting and lowering and individual capability will be reduced.

Does the task involve excessive pushing or pulling of the load?

61   Like lifting, lowering and carrying, the pushing or pulling of a load can be injurious to the handler. The risk of injury is increased if pushing or pulling is carried out with the hands much below knuckle height or above shoulder height.

62   Additionally, because of the way in which pushing and pulling forces have to be transmitted from the handler's feet to the floor, the risk of slipping and consequent injury is much greater. For this reason pushing or pulling a load in circumstances where the grip between foot and floor is poor - whether through the condition of the floor, footwear or both - is likely to increase significantly the risk of injury.

Does the task involve a risk of sudden movement of the load?

63   If a load suddenly becomes free and the handler is unprepared or is not able to retain complete control of the load, unpredictable stresses can be imposed on the body, creating a risk of injury. For example, the freeing of a box jammed on a shelf or the release of a machine component during maintenance work can easily cause injury if handling conditions are not ideal. The risk is compounded if the handler's posture is unstable.

Does the task involve frequent or prolonged physical effort?

64   The frequency with which a load is handled can affect the risk of injury. A quite modest load, handled very frequently, can create as large a risk of

injury as one-off handling of a more substantial load. The effect will be worsened by jerky, hurried movements which can multiply a load's effect on the body.

65   Where physical stresses are prolonged fatigue will occur, increasing the risk of injury. This effect will often be exacerbated by a relatively fixed posture, leading to a rapid increase in fatigue and a corresponding fall in muscular efficiency.

Does the task involve insufficient rest or recovery periods?

66   Research and experience in industry have shown that failure to counter fatigue during physically demanding work increases ill health and reduces output. Consideration should therefore be given to whether there are adequate opportunities for rest (ie breaks from work) or recovery (ie changing to another task which uses a different set of muscles). The amount of work undertaken in fixed postures is also an important consideration since blood flow to the muscles is likely to be reduced, adding to fatigue. This problem is complicated by a large variation in individual susceptibility to fatigue.

Does the task involve a rate of work imposed by a process?

67   Particular care is necessary where the rate of work cannot be varied by the handler. Mild fatigue, which might quickly be relieved by a momentary pause or a brief spell doing another operation using different muscles, can soon become more pronounced, leading to an increased risk of injury.

Handling while seated

68   Handling loads while seated imposes considerable constraints. Use of the relatively powerful leg muscles is precluded and the weight of the handler's body cannot be used as a counterbalance. Therefore most of the work has to be done by the weaker muscles of the arms and trunk.

69   Unless the load is presented close to the body the handler will have to reach and/or lean forward. Not only will handling in this position put the body under additional stress but the seat, unless firmly placed, will then tend to move backwards as the handler attempts to maintain a stable posture.

70   Lifting from below the level of a work surface will almost inevitably result in twisting and stooping, the dangers of which were discussed in paragraphs 55 and 56.

Team handling

71   Handling by two or more people may make possible an operation that is beyond the capability of one person, or reduce the risk of injury to a solo handler. However, team handling may introduce additional problems which the assessment should consider. During the handling operation the proportion of the load that is borne by each member of the team will inevitably vary to some extent. Such variation is likely to be more pronounced on rough ground. Therefore the load that a team can handle in safety is less than the sum of the loads that the individual team members could cope with when working alone.

72   As an approximate guide the capability of a two person team is two thirds the sum of their individual capabilities; and for a three person team the capability is half the sum of their individual capabilities. If steps or slopes must be negotiated most of the weight may be borne by the handler or handlers at the lower end, further reducing the capability of the team as a whole.

Fig 7: Team handling

73   Additional difficulties may arise if team members impede each others' vision or movement, or if the load offers insufficient good handholds. This can occur particularly with compact loads which force the handlers to work close together or where the space available for movement is limited.

The load - making an assessment

Is the load heavy?

74   For many years legislation and guidance on manual handling have concentrated on the weight of the load. It is now well established that the weight of the load is only one - and sometimes not the main - consideration affecting the risk of injury. Other features of the load such as its resistance to movement, its size, shape or rigidity must also be considered. Proper account must also be taken of the circumstances in which the load is handled; for example postural requirements, frequency and duration of handling, workplace design, and aspects of work organisation such as incentive schemes and piecework.

75   Moreover traditional guidance based on so-called 'acceptable' weights has often considered only symmetrical, two-handed lifts, in front of and close to the body. In reality such lifting tasks are comparatively rare since most will involve sideways movement, twisting of the trunk or some other asymmetry. For these reasons an approach to manual handling which concentrates solely upon the weight of the load is likely to be misleading, either failing adequately to deal with the risk of injury or imposing excessively cautious constraints.

76   The numerical guidelines in Appendix 1 consider the weight of the load in relation to other important factors.

Is the load bulky or unwieldy?

77   The shape of a load will affect the way in which it can be held. For example, the risk of injury will be increased if a load to be lifted from the ground is not small enough to pass between the knees, since its bulk will hinder a close approach. Similarly if the bottom front corners of a load are not within

reach when carried at waist height a good grip will be harder to obtain. And if a load to be carried at the side of the body does not clear the ground without requiring the handler to lean away from the load in order to raise it high enough, the handler will be forced into an unfavourable posture.

78   In general if any dimension of the load exceeds about 75 cm its handling is likely to pose an increased risk of injury. This will be especially so if this size is exceeded in more than one dimension. The risk will be further increased if the load does not provide convenient handholds.

79   The bulk of the load can also interfere with vision. Where restriction of view by a bulky load cannot be avoided account should be taken of the increased risk of slipping, tripping, falling or colliding with obstructions.

80   The risk of injury will also be increased if the load is unwieldy and difficult to control. Well-balanced lifting may be difficult to achieve, the load may hit obstructions, or it may be affected by gusts of wind or other sudden air movements.

81   If the centre of gravity of the load is not positioned centrally within the load, inappropriate handling may increase the risk of injury. For example, much of the weight of a typewriter is often at the rear of the machine; therefore an attempt to lift the typewriter from the front will place its centre of gravity further from the handler's body than if the typewriter is first turned around and then lifted from the rear.

82   Sometimes, as with a sealed and unmarked carton, an offset centre of gravity is not visibly apparent. In these circumstances the risk of injury is increased since the handler may unwittingly hold the load with its centre of gravity further from the body than is necessary.

Is the load difficult to grasp?

83   If the load is difficult to grasp, for example because it is large, rounded, smooth, wet or greasy, its handling will call for extra grip strength - which is fatiguing - and will probably entail inadvertent changes of posture. There will also be a greater risk of dropping the load. Handling will be less sure and the risk of injury will be increased.

Is the load unstable, or are its contents likely to shift?

84   If the load is unstable, for example because it lacks rigidity or has contents that are liable to shift, the likelihood of injury is increased. The stresses arising during the manual handling of such a load are less predictable, and the instability may impose sudden additional stresses for which the handler is not prepared. The risks are further increased if the handler is unfamiliar with a particular load and there is no cautionary marking on it.

85   Handling people or animals, for example hospital patients or livestock, can present additional problems. The load lacks rigidity, there is particular concern on the part of the handler to avoid damaging the load, and to complicate matters the load will often have a mind of its own, introducing an extra element of unpredictability. These factors are likely to increase the risk of injury to the handler as compared with the handling of an inanimate load of similar weight and shape.

Is the load sharp, hot or otherwise potentially damaging?

86   Risk of injury may also arise from the external state of the load. It may

have sharp edges or rough surfaces, or be too hot or too cold to touch safely without protective clothing. In addition to the more obvious risk of direct injury, such characteristics may also impair grip, discourage good posture or otherwise interfere with safe handling.

The working environment - making an assessment

Are there space constraints preventing good posture?

87   If the working environment hinders the adoption of good posture the risk of injury from manual handling will be increased. Restricted head room will enforce a stooping posture; furniture, fixtures or other obstructions may increase the need for twisting or leaning; constricted working areas and narrow gangways will hinder the manoeuvring of bulky loads.

Are there uneven, slippery or unstable floors?

88   In addition to increasing the likelihood of slips, trips and falls, uneven or slippery floors hinder smooth movement and create additional unpredictability. Floors which are unstable or susceptible to movement - for example on a boat, a moving train or a mobile work platform - similarly increase the risk of injury through the imposition of sudden, unpredictable stresses.

Are there variations in level of floors or work surfaces?

89   The presence of steps, steep slopes, etc can increase the risk of injury by adding to the complexity of movement when handling loads. Carrying a load up or down a ladder, if it cannot be avoided, is likely to aggravate handling problems because of the additional need to maintain a proper hold on the ladder.

90   Excessive variation between the heights of working surfaces, storage shelving, etc will increase the range of movement and in consequence the scope for injury. This will be especially so if the variation is large and requires, for example, movement of the load from near floor level to shoulder height or beyond.

Are there extremes of temperature or humidity?

91   The risk of injury during manual handling will be increased by extreme thermal conditions. For example, high temperatures or humidity can cause rapid fatigue; and perspiration on the hands may reduce grip. Work at low temperatures may impair dexterity. Gloves and other protective clothing which may be necessary in such circumstances may also hinder movement, impair dexterity and reduce grip. The influence of air movement on working temperatures - the wind chill factor - should not be overlooked.

Are there ventilation problems or gusts of wind?

92   Inadequate ventilation can hasten fatigue, increasing the risk of injury. Sudden air movements, whether caused by a ventilation system or the wind, can make large loads more difficult to manage safely.

Are there poor lighting conditions?

93   Poor lighting conditions can increase the risk of injury. Dimness or glare may cause poor posture, for example by encouraging stooping. Contrast

between areas of bright light and deep shadow can aggravate tripping hazards and hinder the accurate judgement of height and distance.

Individual capability - making an assessment

Does the task require unusual strength, height, etc?

94   Manual handling injuries are more often associated with the nature of the operations than with variations in individual capability. Therefore any assessment which concentrates on individual capability at the expense of task or workplace design is likely to be misleading. However, it is an inescapable fact that the ability to carry out manual handling in safety does vary between individuals.

95   In general the lifting strength of women as a group is less than that of men. But for both men and women the range of individual strength and ability is large, and there is considerable overlap; some women can deal safely with greater loads than some men.

96   An individual's physical capability varies with age, typically climbing until the early 20s, declining gradually during the 40s and more markedly thereafter. It should therefore be recognised that the risk of manual handling injury may be somewhat higher for employees in their teens or in their 50s and 60s, though again the range of individual capability is large and the benefits of experience and maturity should not be overlooked.

97   In deciding whether the physical demands imposed by manual handling operations should be regarded as unusual it is not unreasonable to have some regard to the nature of the work. For example, demands that would be considered unusual for a group of employees engaged in office work might not be regarded as out of the ordinary for a group of employees engaged predominantly in heavy physical labour. It would also be unrealistic to ignore the element of self-selection that often occurs for jobs that are relatively demanding physically.

98   As a general rule, however, the risk of injury should be regarded as unacceptable if the manual handling operations cannot be performed satisfactorily by most reasonably fit, healthy employees.

Does the job put at risk those who might reasonably be considered to be pregnant or to have a health problem?

99   Allowance should be made for pregnancy where the employer could reasonably be expected to be aware of it, ie where the pregnancy is visibly apparent or the employee has informed her employer that she is pregnant. Pregnancy has significant implications for the risk of manual handling injury. Hormonal changes can affect the ligaments, increasing the susceptibility to injury; and postural problems may increase as the pregnancy progresses. Particular care should also be taken for women who may handle loads during the three months following a return to work after childbirth.

100 Allowance should also be made for any health problem of which the employer could reasonably be expected to be aware and which might have a bearing on the ability to carry out manual handling operations in safety. If there is good reason to suspect that an individual's state of health might significantly increase the risk of injury from manual handling operations, medical advice should be sought.

Does the task require special information or training for its safe performance?

101   The risk of injury from a manual handling task will be increased where a worker does not have the information or training necessary for its safe performance. While section 2 of the HSW Act and the Management of Health and Safety at Work Regulations 1992 (see reference section at the back of this publication) require employers to provide safety training, this may need to be supplemented to enable employees to carry out manual handling operations safely.

102   For example, ignorance of any unusual characteristics of loads, or of a system of work designed to ensure safety during manual handling, may lead to injury. Remedial steps such as the provision of mechanical handling aids may themselves create a need for training, for example in the proper use of those aids.

Other factors - making an assessment

Personal protective equipment and other clothing

103   Personal protective equipment should be used only as a last resort, when engineering or other controls do not provide adequate protection. Where the wearing of personal protective equipment cannot be avoided its implications for the risk of manual handling injury should be taken into consideration. For example gloves may impair dexterity; the weight of gas cylinders used with breathing apparatus will increase the stresses on the body. Other clothing such as a uniform required to be worn may inhibit free movement during manual handling.

- (1) Each employer shall -

    (b)   where it is not reasonably practicable to avoid the need for his employees to undertake any manual handling operations at work which involve a risk of their being injured -

        (ii) take appropriate steps to reduce the risk of injury to those employees arising out of their undertaking any such manual handling operations to the lowest level reasonably practicable.

Reducing the risk of injury

Striking a balance

104   It will usually be convenient to continue with the same structured approach used during the assessment of risk, considering in turn the **task**, the **load**, the **working environment** and **individual capability**.

105   The emphasis given to each of these factors may depend in part upon the nature and circumstances of the manual handling operations. Routine manual handling operations carried out in essentially unchanging circumstances, for example in manufacturing processes, may lend themselves particularly to improvement of the task and working environment.

106   However, manual handling operations carried out in circumstances which change continually, for example certain activities carried out in mines or on construction sites, may offer less scope for improvement of the working environment and perhaps the task. More interest may therefore focus on the load - for example can it be made easier to handle?

107   For varied work of this kind, including of course much of the work of the emergency services, the provision of effective training will be especially important. It should enable employees to recognise potentially hazardous handling operations. It should also give them a clear understanding of why they should avoid or modify such operations where possible, make full use of appropriate equipment and apply good handling technique.

An ergonomic approach

108   However, health, safety and productivity are most likely to be optimised if an ergonomic approach is used to design the manual handling operations as a whole. Wherever possible full consideration should be given to the **task**, the **load**, the **working environment**, **individual capability** and the relationship between them, with a view to fitting the operations to the individual rather than the other way around.

109   While better job or workplace design may not eliminate handling injuries, the evidence is that it can greatly reduce them. Particular consideration should be given to the provision of mechanical assistance where this is reasonably practicable.

Mechanical assistance

110   Mechanical assistance involves the use of handling aids - an element of manual handling is retained but bodily forces are applied more efficiently, reducing the risk of injury. There are many examples. A simple lever can reduce the risk of injury merely by lessening the bodily force required to move a load, or by removing fingers from a potentially damaging trap. A hoist, either powered or hand operated, can support the weight of a load and leave the handler free to control its positioning. A trolley, sack truck or roller conveyor can greatly reduce the effort required to move a load horizontally. Chutes are a convenient way of using gravity to move loads from one place to another. Handling devices such as hand-held hooks or suction pads can simplify the problem of handling a load that is difficult to grasp.

Examples of handling aids

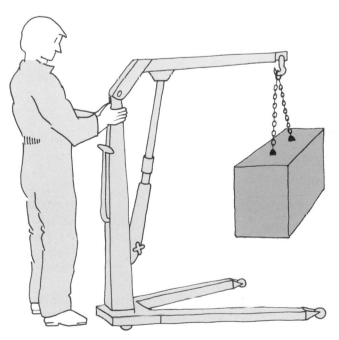

Fig 8: Small hand-powered hydraulic hoist

21

Fig 9: Roller conveyors. Note rotating lift table in background

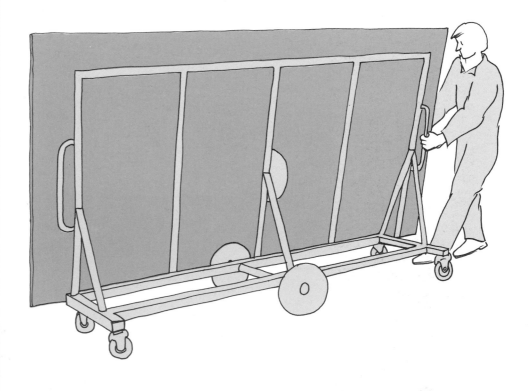

Fig 10: Moving large sheet material

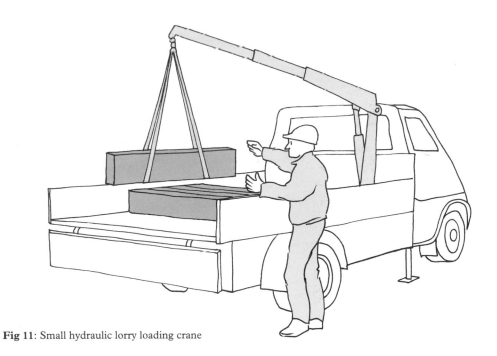

Fig 11: Small hydraulic lorry loading crane

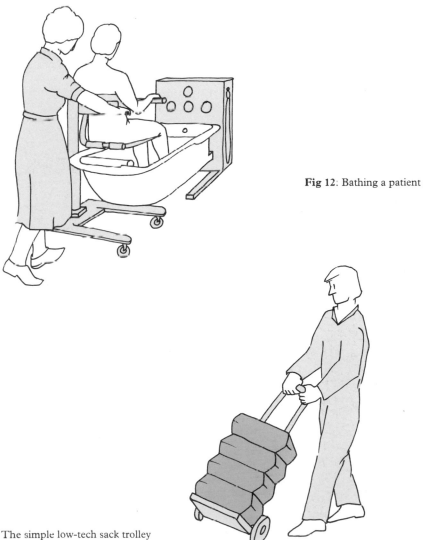

Fig 12: Bathing a patient

Fig 13: The simple low-tech sack trolley

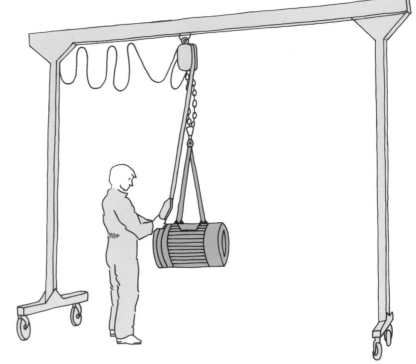

Fig 14: Electric hoist on mobile gantry

Fig 15: Powered vacuum lifter

Involving the workforce

111 Employees, their safety representatives and safety committees should be involved in any redesign of the system of work and encouraged to report its effects. They should be given the opportunity to contribute to the development of good handling practice.

Industry-specific guidance

112 The development of industry-specific guidance within the framework established by the Regulations and this general guidance will provide a valuable source of information on preventive action that has been found effective for particular activities or types of work. Some examples of such guidance are given in the reference section at the back of this document.

'Appropriate' steps

113 Above all, the steps taken to reduce the risk of injury should be 'appropriate'. They should address the problem in a practical and effective manner. Their effectiveness should be monitored; if they do not have the desired effect the situation should be reappraised (see also regulation 4(2) 'reviewing the assessment').

Checklist

114 It may be helpful to use a checklist as an aide-memoire while seeking practical steps to reduce the risk of injury. Appendix 2 offers an example of such a checklist which combines the assessment of risk required by regulation 4(1)(b)(i) with the identification of remedial steps as required by regulation 4(1)(b)(ii). The particular example given will not be suitable in all circumstances; it can be adapted or modified as appropriate.

The task - reducing the risk of injury

Improving task layout

115 Changes to the layout of the task can reduce the risk of injury by, for example, improving the flow of materials or products. Such changes will often bring the additional benefits of increased efficiency and productivity.

116 The optimum position for storage of loads, for example, is around waist height; storage much above or below this height should be reserved for loads that are lighter or more easily handled, or loads that are handled infrequently.

Using the body more efficiently

117 A closely related set of considerations concerns the way in which the handler's body is used. Changes to the task layout, the equipment used, or the sequence of operations can reduce or remove the need for twisting, stooping and stretching.

118 In general, any change that allows the load to be held closer to the body is likely to reduce the risk of injury. The level of stress at the lower back will be reduced; the weight of the load will be more easily counterbalanced by the weight of the body; and the load will be more stable and the handler less likely to lose control of it. Moreover, if the load is hugged to the body friction with the handler's garments will steady it and may help to support its weight. The

need for protective clothing should be considered (see paragraphs 131 and 132).

119 When the lifting of loads at or near floor level is unavoidable, handling techniques which allow the use of the relatively strong leg muscles rather than those of the back are preferable provided the load is small enough to be held close to the trunk.

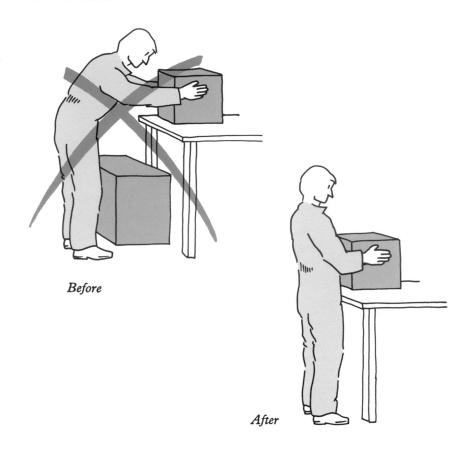

Before

After

Fig 16: Avoiding an obstructed lift. Organise the workplace so that the handler can get as close to the load as possible

120 The closeness of the load to the body can also be influenced by foot placement. The elimination of obstacles which need to be reached over or into - for example poorly placed pallets, excessively deep bins - will permit the handler's feet to be placed beneath or adjacent to the load.

121 Where possible the handler should be able to move in close to the load before beginning the manual handling operation. The handler should also be able to address the load squarely, preferably facing in the direction of intended movement.

122 The risk of injury may also be reduced if lifting can be replaced by controlled pushing or pulling. For example it may be possible to slide the load or roll it along. However, uncontrolled sliding or rolling, particularly of large or heavy loads, may introduce fresh risks of injury.

123 For both pulling and pushing, a secure footing should be ensured, and the hands applied to the load at a height between waist and shoulder wherever possible. A further option, where other safety considerations allow, is to push with the handler's back against the load, using the strong leg muscles to exert the force.

Fig 17: Hand position when pushing

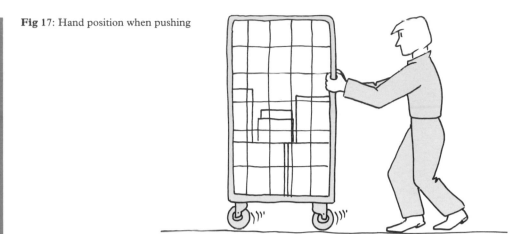

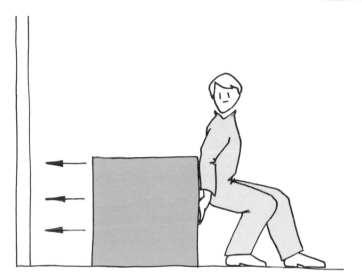

Fig 18: Using the strong leg muscles

Improving the work routine

124   The risk of manual handling injury can also be reduced by careful attention to the work routine. Minimising the need for fixed postures dictated by sustained holding or supporting of a load will reduce fatigue and the associated fall-off in muscular efficiency. Attention to the frequency of handling loads, especially those that are heavy or awkward, can also reduce fatigue and the risk of injury. Where possible, tasks should be self-paced and employees trained to adjust their rate of working to optimise safety and productivity.

125   An inflexible provision of rest pauses may not be an efficient method of reducing the risk of injury. The large variation in individual susceptibility to muscular fatigue means that mandatory, fixed breaks are generally less effective than those taken voluntarily within the constraints of what is possible in terms of work organisation.

126   A better solution can often be found in job rotation where this allows one group of muscles to rest while others are being used. Periods of heavy work may be interspersed with lighter activities such as paper work or the monitoring of instruments. Job rotation can also bring advantages in reduced monotony and increased attentiveness. However, where rotation merely repeats the use of the same group of muscles, albeit on a different task, it is generally ineffective in reducing the risk of manual handling injury.

Handling while seated

127 For the reasons given in paragraphs 68 to 70 the loads that can be handled in safety by a person who is seated are substantially less than can be dealt with while standing. This activity therefore demands particular care. Lifting loads from the floor while seated should be avoided where possible.

128 The possibility of accidental movement of the seat should be considered. Castors may be inadvisable, especially on hard floors. A swivel-action seat will help the handler to face the load without having to twist the trunk. The relative heights of seats and work surfaces should be well matched. Further advice on this is given in the HSE booklet *Seating at work* (see reference section).

Team handling

129 Where a handling operation would be difficult or unsafe for one person, handling by a team of two or more may provide an answer. However, team handling can introduce additional hazards and caution should be exercised for the reasons given in paragraphs 71 to 73.

130 For safe team handling there should be enough space for the handlers to manoeuvre as a group. They should have adequate access to the load, and the load should provide sufficient handholds; if the load is particularly small or difficult to grasp then a handling aid such as a stretcher or slings should be used. One person should plan and then take charge of the operation, ensuring that movements are coordinated. Team members should preferably be of broadly similar height and physical capability.

Personal protective equipment and other clothing

131 The nature of the load, or the environment in which it is handled, may necessitate the use of personal protective equipment (PPE) such as gloves, aprons, overalls, gaiters or safety footwear. In these cases the protection offered by PPE should not be compromised to facilitate the manual handling operations. Alternative methods of handling may need to be considered where manual handling is likely to lead to risks from the contents of the load or from external contamination.

132 PPE and indeed all work clothing should be well fitting and restrict movement as little as possible. Fasteners, pockets and other features on which loads might snag should be concealed. Gloves should be close fitting and supple, to interfere with manual dexterity as little as possible. Footwear should provide adequate support, a stable, non-slip base and proper protection. Restrictions on the handler's movement caused by wearing protective clothing need to be recognised in the design of the task. Reference should be made to The Personal Protective Equipment at Work Regulations 1992 (see reference section).

Maintenance and accessibility of equipment

133 All equipment provided for use during manual handling, including handling aids and PPE, should be well maintained and there should be a defect reporting and correction system. The siting of equipment can be important: handling aids and PPE that are not readily accessible are less likely to be used fully and effectively. Reference should be made to The Provision and Use of Work Equipment Regulations 1992 (see reference section).

Safety of machinery - European standards

134   Under the Machinery Directive 89/392/EEC (as amended) machinery placed on the market must satisfy certain essential health and safety requirements. This is a product safety Directive made under Article 100A of the EC Treaty and will be implemented by the Department of Trade and Industry. One of the Directive's requirements is that machinery be designed to facilitate its safe handling.

135   In support of this requirement CEN, one of the European standards making bodies, is preparing a harmonised standard entitled *Safety of machinery - human physical performance*. The purpose of the proposed standard is to assist machinery designers and manufacturers. It has no status in relation to Directive 90/269/EEC on the manual handling of loads which was made under Article 118A of the Treaty. The standard will therefore have no direct relevance to the Manual Handling Operations Regulations 1992.

The load - reducing the risk of injury

Making it lighter

136   Where a risk of injury from the manual handling of a load is identified, consideration should be given to reducing its weight. For example liquids and powders may be packaged in smaller containers. Where loads are bought in it may be possible to specify lower package weights. However the breaking down of loads will not always be the safest course: the consequent increase in the frequency of handling should not be overlooked.

137   If a great variety of weights is to be handled it may be possible to sort the loads into weight categories so that additional precautions can be applied selectively, where most needed.

Making it smaller or easier to manage

138   Similarly, consideration should be given to making loads less bulky so that they can be grasped more easily and the centre of gravity brought closer to the handler's body. Again, it may be possible to specify smaller or more manageable loads, or to redesign those produced in-house.

Making it easier to grasp

139   Where the size, surface texture or nature of a load makes it difficult to grasp, consideration should be given to the provision of handles, hand grips, indents or any other feature designed to improve the handler's grasp. Alternatively it may be possible to place the load securely in a container which is itself easier to grasp. Where a load is bulky rather than heavy it may be easier to carry it at the side of the body if it has suitable handholds or if slings or other carrying devices can be provided.

140   The positioning of handholds can play a part in reducing the risk of injury. For example, handholds at the top of a load may reduce the temptation to stoop when lifting it from a low level. However, depending upon the size of the load, this might also necessitate carriage with bent arms which could increase fatigue.

141   Handholds should be wide enough to clear the breadth of the palm, and deep enough to accommodate the knuckles and any gloves which may need to be worn.

Making it more stable

142   Where possible, packaging should be such that objects will not shift unexpectedly while being handled. Where the load as a whole lacks rigidity it may be preferable to use slings or other aids to maintain effective control during handling. Ideally, containers holding liquids or free-moving powders should be well filled, leaving only a small amount of free space; where this is not possible alternative means of handling should be considered.

Making it less damaging to hold

143   As far as possible loads should be clean and free from dust, oil, corrosive deposits, etc. To prevent injury during the manual handling of hot or cold materials an adequately insulated container should be used; failing this suitable handling aids or PPE will be necessary. Sharp corners, jagged edges, rough surfaces etc should be avoided where possible; again, where this cannot be achieved then the use of handling aids or PPE will be necessary. In selecting personal protective equipment the advice given in paragraphs 131 and 132 should be noted.

The working environment - reducing the risk of injury

Removing space constraints

144   Gangways and other working areas should where possible allow adequate room to manoeuvre during manual handling operations. The provision of sufficient clear floor space and head room is important; constrictions caused by narrow doorways and the positioning of fixtures, machines, etc should be avoided as far as possible. In many cases much can be achieved simply by improving the standard of housekeeping. Reference should be made to The Workplace (Health, Safety and Welfare) Regulations 1992 (see reference section at the back of this publication).

The nature and condition of floors

145   On permanent sites, both indoors and out, a flat, well maintained and properly drained surface should be provided. In construction, agriculture and other activities where manual handling may take place on temporary surfaces, the ground should be prepared if possible and kept even and firm; if possible suitable coverings should be provided. Temporary work platforms should be firm and stable.

146   Spillages of water, oil, soap, food scraps and other substances likely to make the floor slippery should be cleared away promptly. Where necessary, and especially where floors can become wet, attention should be given to the choice of slip-resistant surfacing.

147   Particular care is necessary where manual handling is carried out on a surface that is unstable or susceptible to movement, as for example on a boat, a moving train or a mobile work platform. In these conditions the capability to handle loads in safety may be reduced significantly.

Working at different levels

148   Where possible all manual handling activities should be carried out on a single level. Where more than one level is involved the transition should preferably be made by a gentle slope or, failing that, by well positioned and properly maintained steps. Manual handling on steep slopes should be avoided

as far as possible. Working surfaces such as benches should, where possible, be at a uniform height to reduce the need for raising or lowering loads.

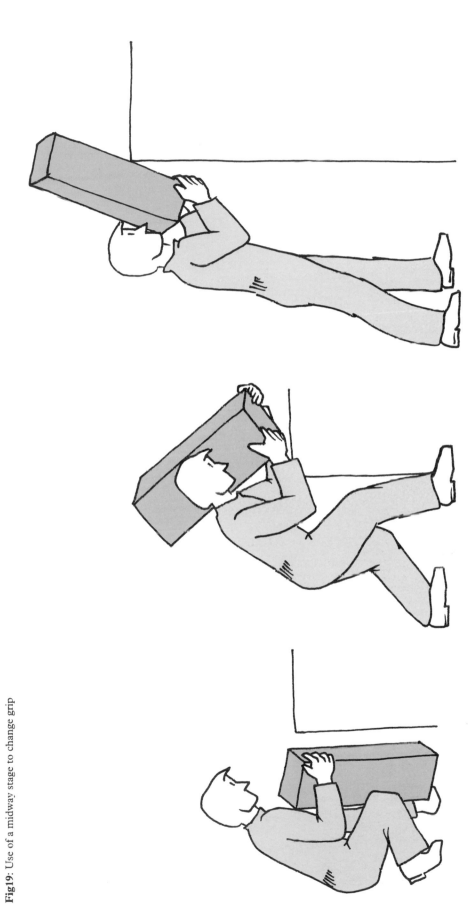

Fig 19: Use of a midway stage to change grip

The thermal environment and ventilation

149 There is less risk of injury if manual handling is performed in a comfortable working environment. Extremes of temperature, excessive humidity and poor ventilation should be avoided where possible, either by improving environmental control or relocating the work.

150 Where these conditions cannot be changed, for example when manual handling is necessarily performed out of doors in extreme weather, or close to a very hot process, or in a refrigerated storage area, and the use of PPE is necessary, the advice given in paragraphs 131 and 132 should be noted.

Strong air movements

151 Particular care should be taken when handling bulky or unwieldy loads in circumstances in which high winds or powerful ventilation systems could catch a load and destabilise the handler. Possible improvements include relocating the handling operations or taking a different route, provision of handling aids to give greater control of the load, or team handling.

Lighting

152 There should be sufficient well-directed light to enable handlers to see clearly what they are doing and the layout of the workplace, and to make accurate judgements of distance and position.

Individual capability - reducing the risk of injury

Personal considerations

153 Particular consideration should be given to employees who are or have recently been pregnant, or who are known to have a history of back trouble, hernia or other health problems which could affect their manual handling capability. However, beyond such specific pointers to increased risk of injury the scope for preventive action on an individual basis is limited.

154 Clearly an individual's state of health, fitness and strength can significantly affect the ability to perform a task safely. But even though these characteristics vary enormously, studies have shown no close correlation between any of them and injury incidence. There is therefore insufficient evidence for reliable selection of individuals for safe manual handling on the basis of such criteria. It is recognised, however, that there is often a degree of self-selection for work that is physically demanding.

155 It is also recognised that motivation and self-confidence in the ability to handle loads are important factors in reducing the risk of injury. These are linked with fitness and familiarity. Unaccustomed exertion - whether in a new task or on return from holiday or sickness absence - can carry a significant risk of injury and requires particular care.

Information and training

156 Section 2 of the HSW Act and regulations 8 and 11 of the Management of Health and Safety at Work Regulations 1992 require employers to provide their employees with health and safety information and training. This should be supplemented as necessary with more specific information and training on manual handling injury risks and prevention, as part of the steps to reduce risk required by regulation 4(1)(b)(ii) of the present Regulations.

157　It should not be assumed that the provision of information and training alone will ensure safe manual handling. The primary objective in reducing the risk of injury should always be to optimise the design of the manual handling operations, improving the task, the load and the working environment as appropriate. Where possible the manual handling operations should be designed to suit individuals, not the other way round. However as a complement to a safe system of work, rather than a substitute for it, effective training has an important part to play in reducing the risk of manual handling injury.

158　Employers should ensure that their employees understand clearly how manual handling operations have been designed to ensure their safety. Employees, their safety representatives and safety committees should be involved in the development and implementation of manual handling training, and the monitoring of its effectiveness.

159　In devising a training programme for safe manual handling, particular attention should therefore be given to imparting a clear understanding of:

(a)　how potentially hazardous handling operations may be recognised;

(b)　how to deal with unfamiliar handling operations;

(c)　the proper use of handling aids;

(d)　the proper use of personal protective equipment;

(e)　features of the working environment that contribute to safety;

(f)　the importance of good housekeeping;

(g)　factors affecting individual capability;

(h)　good handling technique (see paragraphs 162 to 164).

160　Employees should be trained to recognise loads whose weight, in conjunction with their shape and other features, and the circumstances in which they are handled, might cause injury. Simple methods for estimating weight on the basis of volume may be taught. Where volume is less important than the density of the contents, as for example in the case of a dustbin containing refuse, an alternative technique for assessing the safety of handling should be taught, such as rocking the load from side to side before attempting to lift it.

161　In general, unfamiliar loads should be treated with caution. For example, it should not be assumed that apparently empty drums or other closed containers are in fact empty. The load may first be tested, for example by attempting to raise one end. Employees should be taught to apply force gradually until either undue strain is felt, in which case the task should be reconsidered, or it is apparent that the task is within the handler's capability.

Fig 20: Rocking a load to assess its ease of handling

Good handling technique

162   The development of good handling technique is no substitute for other risk reduction steps such as improvements to the task, load or working environment, but it will form a very valuable adjunct to them.   It requires both training and practice.   The training should be carried out in conditions that are as realistic as possible, thereby emphasising its relevance to everyday handling operations.

163   The content of training in good handling technique should be tailored to the particular handling operations likely to be undertaken.   It should begin with relatively simple examples and progress to more specialised handling operations as appropriate.   The following list illustrates some important points, using a basic lifting operation by way of example (Fig 21):

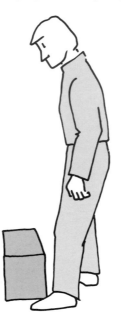

Stop and think. Plan the lift. Where is the load going to be placed? Use appropriate handling aids if possible. Do you need help with the load? Remove obstructions such as discarded wrapping materials. For a long lift - such as floor to shoulder height - consider resting the load mid-way on a table or bench in order to change grip.

Place the feet. Feet apart, giving a balanced and stable base for lifting (tight skirts and unsuitable footwear made this difficult). Leading leg as far forward as is comfortable.

Adopt a good posture. Bend the knees so that the hands when grasping the load are as nearly level with the waist as possible. But do not kneel or overflex the knees. Keep the back straight (tucking in the chin helps). Lean forward a little over the load if necessary to get a good grip. Keep shoulders level and facing in the same direction as the hips.

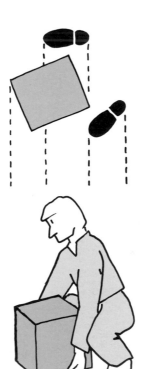

Get a firm grip. Try to keep the arms within the boundary formed by the legs. The optimum position and nature of the grip depends on the circumstances and individual preference, but it must be secure. A hook grip is less fatiguing than keeping the fingers straight. If it is necessary to vary the grip as the lift proceeds, do this as smoothly as possible.

Don't jerk. Carry out the lifting movement smoothly, keeping control of the load.

Move the feet. Don't twist the trunk when turning to the side.

Keep close to the load. Keep the load close to the trunk for as long as possible. Keep the heaviest side of the load next to the trunk. If a close approach to the load is not possible try sliding it towards you before attempting to lift it.

Put down, then adjust. If precise positioning of the load is necessary, put it down first, then slide it into the desired position.

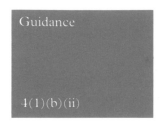
Vocational qualifications

164   The development of specific statements of what needs to be done, how well and by whom (ie statements of competence) will help to determine the extent of any shortfall in training. Such statements may be embodied in qualifications accredited by the National Council for Vocational Qualifications (NCVQ) and SCOTVEC (the Scottish Vocational Education Council).

Regulation

4(1)(b)(iii)

- (1) Each employer shall -

(b)   *where it is not reasonably practicable to avoid the need for his employees to undertake any manual handling operations at work which involve a risk of their being injured -*

(iii)   *take appropriate steps to provide any of those employees who are undertaking any such manual handling operations with general indications and, where it is reasonably practicable to do so, precise information on -*

(aa)   *the weight of each load, and*

(bb)   *the heaviest side of any load whose centre of gravity is no positioned centrally.*

The load - providing additional information

165   Regulation 4(1)(b)(iii) can be satisfied in a variety of ways, depending upon the circumstances.

166   The requirement to provide 'general indications' of the weight and nature of the loads to be handled should be addressed during basic training so that employees are suitably prepared for the operations they are likely to undertake.

167   In addition, where it is reasonably practicable to do so, employers should give more precise information. Employers whose businesses originate loads may find that this information is best given by marking it on the loads.

168   The Regulations impose duties on employers whose employees carry out manual handling. However, those who originate loads - manufacturers, packers, etc - that are likely to undergo manual handling may also have relevant duties, for example under sections 3 or 6 of the HSW Act, for the health and safety of other people at work. They should give particular consideration to making loads easy to grasp and handle and to marking loads clearly with their weight and, where appropriate, an indication of their heaviest side.

Regulation

4(2)

- (2) Any assessment such as is referred to in paragraph (1)(b)(i) of this regulation shall be reviewed by the employer who made it if -

(a)   *there is reason to suspect that it is no longer valid; or*

(b)   *there has been a significant change in the manual handling operations to which it relates;*

and where as a result of any such review changes to an assessment are required, the relevant employer shall make them.

Reviewing the assessment

169 The assessment should be kept up-to-date. It should be reviewed if new information comes to light or if there has been a change in the manual handling operations which, in either case, could materially have affected the conclusion reached previously. The assessment should also be reviewed if a reportable injury occurs. It should be corrected or modified where this is found to be necessary.

Regulation 5

Duty of employees

Regulation
5

Each employee while at work shall make full and proper use of any system of work provided for his use by his employer in compliance with regulation 4(1)(b)(ii) of these Regulations.

170 Duties are already placed on employees by section 7 of the HSW Act, under which they must:

(a) take reasonable care for their own health and safety and that of others who may be affected by their activities; and

(b) cooperate with their employers to enable them to comply with their health and safety duties.

171 In addition, regulation 12 of the Management of Health and Safety at Work Regulations 1992 requires employees generally to make use of appropriate equipment provided for them, in accordance with their training and the instructions their employer has given them. Such equipment will include machinery and other aids provided for the safe handling of loads.

172 Regulation 5 of the present Regulations supplements these general duties in the case of manual handling by requiring employees to follow appropriate systems of work laid down by their employer to promote safety during the handling of loads.

Emergency action

173 These provisions do not preclude well-intentioned improvisation in an emergency, for example during efforts to rescue a casualty, fight a fire or contain a dangerous spillage.

Regulation 6

Exemption certificates

- (1) The Secretary of State for Defence may, in the interests of national security, by a certificate in writing exempt -

(a) any of the home forces, any visiting force or any headquarters from any requirement imposed by regulation 4 of these Regulations; or

(b) any member of the home forces, any member of a visiting force or any member of a headquarters from the requirement imposed by regulation 5 of these Regulations;

and any exemption such as is specified in sub-paragraph (a) or (b) of this paragraph may be granted subject to conditions and to a limit of time and may be revoked by the

37

said Secretary of State by a further certificate in writing at any time.

(2) In this regulation –

(a) *"the home forces" has the same meaning as in section 12(1) of the Visiting Forces Act 1952(a);*

(b) *"headquarters" has the same meaning as in article 3(2) of the Visiting Forces and International Headquarters (Application of Law) Order 1965(b);*

(c) *"member of a headquarters" has the same meaning as in paragraph 1(1) of the Schedule to the International Headquarters and Defence Organisations Act 1964(c); and*

(d) *"visiting force" has the same meaning as it does for the purposes of any provision of Part I of the Visiting Forces Act 1952.*

(a)   1952 c.67
(b)   SI 1965/1536, to which these are amendments not relevant to these Regulations.
(c)   1964 c.5

Regulation 7

Extension outside Great Britain

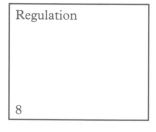

These Regulations shall, subject to regulation 3 hereof, apply to and in relation to the premises and activities outside Great Britain to which sections 1 to 59 and 80 to 82 of the Health and Safety at Work etc Act 1974 apply by virtue of the Health and Safety at Work etc Act 1974 (Application Outside Great Britain) Order 1989(d) as they apply within Great Britain.

(d)   SI 1989/840

174   The Regulations apply to offshore activities covered by the 1989 Order on or associated with oil and gas installations, including mobile installations, diving support vessels, heavy lift barges and pipe-lay barges.

Regulation 8

Repeals and revocations

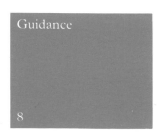

- (1) The enactments mentioned in column 1 of Part I of Schedule 2 to these Regulations are repealed to the extent specified in the corresponding entry in column 3 of that part.

(2) The Regulations mentioned in column 1 of Part II of Schedule 2 to these Regulations are revoked to the extent specified in the corresponding entry in column 3 of that part.

175   The Regulations, like the European Directive on manual handling, apply a modern ergonomic approach to the prevention of injury: they take account of a wide range of relevant factors including the nature of the task, the load, the working environment and individual capability. The Regulations have therefore replaced a number of outdated provisions which concentrated on the weight of the load being handled. The provisions are listed in Schedule 2 to the Regulations.

Factors to which the employer must have regard and questions he must consider when making an assessment of manual handling operations

Regulation 4(1)(b)(i)

Column 1 *Factors*	Column 2 *Questions*

1 The tasks	Do they involve:
	
	- holding or manipulating loads at distance from trunk?
	- unsatisfactory bodily movement or posture, especially:
	  - twisting the trunk?
	  - stooping?
	  - reaching upwards?
	- excessive movement of loads, especially:
	  - excessive lifting or lowering distances?
	  - excessive carrying distances?
	- excessive pushing or pulling of loads?
	- risk of sudden movement of loads?
	- frequent or prolonged physical effort?
	- insufficient rest or recovery periods?
	- a rate of work imposed by a process?

2 The loads	Are they: - heavy? - bulky or unwieldy? - difficult to grasp? - unstable, or with contents likely to shift? - sharp, hot or otherwise potentially damaging?
3 The working environment	Are there: - space constraints preventing good posture? - uneven, slippery or unstable floors? - variations in level of floors or work surfaces? - extremes of temperature or humidity? - conditions causing ventilation problems or gusts of wind? - poor lighting conditions?

4 Individual capability

Does the job:

- require unusual strength, height, etc?

- create a hazard to those who might reasonably be considered to be pregnant or to have a health problem?

- require special information or training for its safe performance?

5 Other factors

Is movement or posture hindered by personal protective equipment or by clothing?

Schedule 2

Repeals and revocations

Regulation 8

Part I Repeals

Column 1 *Short title of enactment*	Column 2 *Reference*	Column 3 *Extent of repeal*
The Children and Young Persons Act 1933.	1933 c.12	Section 18(1)(f) except insofar as that paragraph applies to such employment as is permitted under section 1(2) of the Employment of Women, Young Persons, and Children Act 1920 (1920 c. 65).
The Children and Young Persons (Scotland) Act 1937.	1937 c.37	Section 28(1)(f) except insofar as that paragraph applies to such employment as is permitted under section 1(2) of the Employment of Women, Young Persons, and Children Act 1920.
The Mines and Quarries Act 1954.	1954 c.70	Section 93; in section 115 the word "ninety-three".
The Agriculture (Safety, Health and Welfare Provisions) Act 1956.	1956 c.49	Section 2.
The Factories Act 1961.	1961 c.34	Section 72.
The Offices, Shops and Railway Premises Act 1963.	1963 c.41	Section 23 except insofar as the prohibition contained in that section applies to any person specified in section 90(4) of the same Act; in section 83(1) the number "23".

Part II Revocations

Column 1 *Title of instrument*	Column 2 *Reference*	Column 3 *Extent of revocation*
The Agriculture (Lifting of Heavy Weights) Regulations 1959.	SI 1959/2120	The whole Regulations.
The Construction (General Provisions) Regulations 1961.	SI 1961/1580	In regulation 3(1)(a) the phrase "and 55"; regulation 55.

Appendix 1   Numerical guidelines for assessment

Introduction - the need for assessment

1   Regulation 3(1) of the Management of Health and Safety at Work Regulations 1992 (see Reference section at the back of this publication) requires employers to make a suitable and sufficient assessment of the risks to the health and safety of their employees while at work. Where this general assessment indicates the possibility of risks to employees from the manual handling of loads the requirements of the Manual Handling Operations Regulations 1992 (the Regulations) should be considered.

2   Regulation 4(1) of the Regulations sets out a hierarchy of measures for safety during manual handling:

(a)   avoid hazardous manual handling operations so far as is reasonably practicable;

(b)   make a suitable and sufficient assessment of any hazardous manual handling operations that cannot be avoided; and

(c)   reduce the risk of injury from those operations so far as is reasonably practicable.

Purpose of the guidelines

3   The Manual Handling Operations Regulations, like the European Directive on manual handling, set no specific requirements such as weight limits. Instead, assessment based on a range of relevant factors listed in Schedule 1 to the Regulations is used to determine the risk of injury and point the way to remedial action. However a full assessment of every manual handling operation could be a major undertaking and might involve wasted effort.

4   The following numerical guidelines therefore provide an initial filter which can help to identify those manual handling operations deserving more detailed examination. The guidelines set out an approximate boundary within which operations are unlikely to create a risk of injury sufficient to warrant more detailed assessment. This should enable assessment work to be concentrated where it is most needed.

5   There is no threshold below which manual handling operations may be regarded as 'safe'. Even operations lying within the boundary mapped out by the guidelines should be avoided or made less demanding wherever it is reasonably practicable to do so.

Source of the guidelines

6   These guidelines have been drawn up by HSE's medical and ergonomics experts on the basis of a careful study of the published literature and their own extensive practical experience of assessing risks from manual handling operations.

Individual capability

7   There is a wide range of individual physical capability, even among those fit and healthy enough to be at work. For the working population the guideline figures will give reasonable protection to nearly all men and between one half and two thirds of women. To provide the same degree of protection to nearly

all working women the guideline figures should be reduced by about one third. 'Nearly all' in this context means about 95%.

8   It is important to understand that the **guideline figures are not limits.** They may be exceeded where a more detailed assessment shows that it is appropriate to do so, having regard always to the employer's duty to avoid or reduce risk of injury where this is reasonably practicable. However, even for a minority of fit, well-trained individuals working under favourable conditions any operations which would exceed the guideline figures by more than a factor of about two should come under very close scrutiny.

Guidelines for lifting and lowering

9   Basic guideline figures for manual handling operations involving lifting and lowering are set out in Figure 1. They assume that the load is readily grasped with both hands and that the operation takes place in reasonable working conditions with the handler in a stable body position.

10   The guideline figures take into consideration the vertical and horizontal position of the hands as they move the load during the handling operation, as well as the height and reach of the individual handler. It will be apparent that the capability to lift or lower is reduced significantly if, for example, the load is held at arm's length or the hands pass above shoulder height.

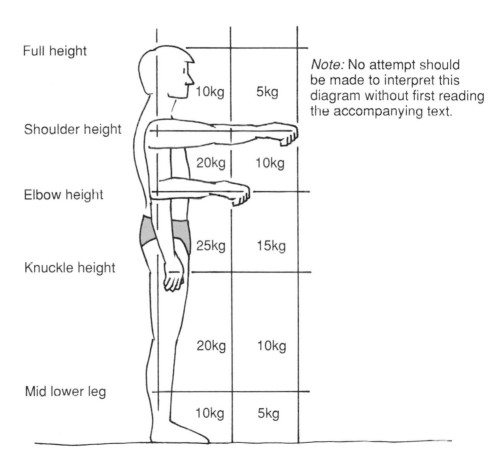

Fig 1: Lifting and lowering

11   If the hands enter more than one of the box zones during the operation the smallest weight figure should be used. The transition from one box zone to another is not abrupt; an intermediate figure may be chosen where the hands are close to a boundary. Where lifting or lowering with the hands beyond the box zones is unavoidable a more detailed assessment should be made.

Twisting

12   The basic guideline figures for lifting and lowering should be reduced if the handler twists to the side during the operation. As a rough guide the figures should be reduced by about 10% where the handler twists through 45° and by about 20% where the handler twists through 90°.

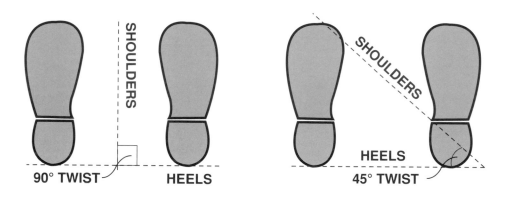

Fig 2: Assessing twist

Frequent lifting and lowering

13   The basic guideline figures for lifting and lowering are for relatively infrequent operations - up to approximately 30 operations per hour - where the pace of work is not forced, adequate pauses for rest or recovery are possible and the load is not supported for any length of time. They should be reduced if the operation is repeated more frequently. As a rough guide the figures should be reduced by 30% where the operation is repeated once or twice per minute, by 50% where the operation is repeated around five to eight times per minute and by 80% where the operation is repeated more than about 12 times per minute.

Guidelines for carrying

14   Basic guideline figures for manual handling operations involving carrying are similar to those given for lifting and lowering, though carrying will not normally be carried out with the hands below knuckle height.

15   It is also assumed that the load is held against the body and is carried no further than about 10 m without resting. If the load is carried over a longer distance without resting the guideline figures may need to be reduced.

16   Where the load can be carried securely on the shoulder without first having to be lifted (as for example when unloading sacks from a lorry) a more detailed assessment may show that it is acceptable to exceed the guideline figure.

Guidelines for pushing and pulling

17   The following guideline figures are for manual handling operations involving pushing and pulling, whether the load is slid, rolled or supported on wheels. The guideline figure for starting or stopping the load is a force of about 25 kg (ie about 250 Newtons). The guideline figure for keeping the load in motion is a force of about 10 kg (ie about 100 Newtons).

Fig 3: Measuring pulling force

18   It is assumed that the force is applied with the hands between knuckle and shoulder height; if this is not possible the guideline figures may need to be reduced. No specific limit is intended as to the distance over which the load is pushed or pulled provided there are adequate opportunities for rest or recovery.

Guidelines for handling while seated

19   The basic guideline figure for handling operations carried out while seated is given in Figure 4 and applies only when the hands are within the box zone indicated. If handling beyond the box zone is unavoidable or, for example, there is significant twisting to the side a more detailed assessment should be made.

Fig 4: Handling while seated

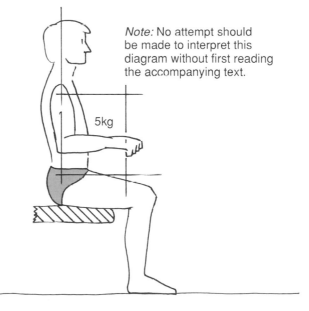

Note: No attempt should be made to interpret this diagram without first reading the accompanying text.

5kg

REMEMBER - the guideline figures should not be regarded as precise recommendations. They should be applied with caution. Where doubt remains, a more detailed assessment should be made.

Manual handling of loads

EXAMPLE OF AN ASSESSMENT CHECKLIST

Note: This checklist may be copied freely. It will remind you of the main points to think about while you:
- consider the risk of injury from manual handling operations
- identify steps that can remove or reduce the risk
- decide your priorities for action.

SUMMARY OF ASSESSMENT	Overall priority for remedial action: Nil / Low / Med / High*
Operations covered by this assessment:	Remedial action to be taken: ....
....	....
....	....
Locations:	Date by which action is to be taken: ....
Personnel involved:	Date for reassessment: ....
Date of assessment:	Assessor's name: .... Signature: ....

*circle as appropriate

Section A - Preliminary:

Q1   **Do the operations involve a significant risk of injury?**   Yes / No*

         If 'Yes' go to Q2. If 'No' the assessment need go no further.

         If in doubt answer 'Yes'. You may find the guidelines in Appendix 1 helpful.

Q2   **Can the operations be avoided / mechanised / automated at reasonable cost?**   Yes / No*

         If 'No' go to Q3. If 'Yes' proceed and then check that the result is satisfactory.

Q3   **Are the operations clearly within the guidelines in Appendix 1?**   Yes / No*

         If 'No' go to Section B. If 'Yes' you may go straight to Section C if you wish.

Section C - Overall assessment of risk:

Q   **What is your overall assessment of the risk of injury?**   **Insignificant / Low / Med / High***

         If not 'Insignificant' go to Section D. If 'Insignificant' the assessment need go no further.

Section D - Remedial action:

Q   **What remedial steps should be taken, in order of priority?**

     i ....

     ii ....

     iii ....

     iv ....

     v ....

And finally:

     - complete the SUMMARY above

     - compare it with your other manual handling assessments

     - decide your priorities for action

     - TAKE ACTION.................AND CHECK THAT IT HAS THE DESIRED EFFECT

Section B - More detailed assessment, where necessary:

Questions to consider: (If the answer to a question is 'Yes' place a tick against it and then consider the level of risk)	Yes	Level of risk: (Tick as appropriate)			Possible remedial action: (Make rough notes in this column in preparation for completing Section D)
		Low	Med	High	
The tasks - do they involve:					
◆ holding loads away from trunk?					
◆ twisting?					
◆ stooping?					
◆ reaching upwards?					
◆ large vertical movement?					
◆ long carrying distances?					
◆ strenuous pushing or pulling?					
◆ unpredictable movement of loads?					
◆ repetitive handling?					
◆ insufficient rest or recovery?					
◆ a workrate imposed by a process?					
The loads - are they:					
◆ heavy?					
◆ bulky/unwieldy?					
◆ difficult to grasp?					
◆ unstable/unpredictable?					
◆ intrinsically harmful (eg sharp/hot?)					
The working environment - are there:					
◆ constraints on posture?					
◆ poor floors?					
◆ variations in levels?					
◆ hot/cold/humid conditions?					
◆ strong air movements?					
◆ poor lighting conditions?					
Individual capability - does the job:					
◆ require unusual capability?					
◆ hazard those with a health problem?					
◆ hazard those who are pregnant?					
◆ call for special information/training?					
Other factors - Is movement or posture hindered by clothing or personal protective equipment?					

Deciding the level of risk will inevitably call for judgement. The guidelines in Appendix 1 may provide a useful yardstick.

When you have completed Section B go to Section C.

References and further information

The EC Directive on manual handling
Council Directive of 29 May 1990 on the minimum health and safety requirements for the manual handling of loads where there is a risk particularly of back injury to workers (fourth individual Directive within the meaning of Article 16(1) of Directive 89/391/EEC) (90/269/EEC) Official Journal of the European Communities, 21.6.90, Vol 33 No L156 9-13

HSE publications (available from HMSO)
Troup JDG and Edwards FC *Manual handling - a review paper* HMSO ISBN 0 11 883778 8
HSE *Human factors in industrial safety* HS(G)48 *HMSO 1989* ISBN 0 11 885486 0
HSE *Lighting at work* HS(G)38 HMSO 1987 ISBN 0 11 883964 0
HSE *Seating at work* HS(G)57 HMSO 1991 ISBN 0 11 885431 0
HSE *Watch your step - prevention of slipping, tripping and falling accidents at work* HMSO 1985 ISBN 0 11 883782 6
HSE *Work related upper limb disorders - a guide to prevention* HS(G)60 HMSO 1990 ISBN 0 11 885565 4

HSE leaflets (available free from HSE public enquiry points)
HSE *Ergonomics at work* IND(G)90(L) 1990

Other publications
Ayoub MM and Mital A *Manual Materials Handling* 1989 Taylor and Francis ISBN 0 85066 383 0
Galer IAR *Applied Ergonomics Handbook* 2nd edition 1987 Butterworths ISBN 0 40 800800 6
Grandjean E *Fitting the task to the man: a textbook of occupational ergonomics* 4th edition 1988 Taylor and Francis ISBN 0 85 066379 2
Nicholson A S and Ridd J (eds) *Health, safety and ergonomics* 1988 Butterworths ISBN 0 40 802386 4
Pheasant S *Bodyspace: anthropometry, ergonomics and design* 1986 Taylor and Francis ISBN 0 85 066352 0
Pheasant S *Ergonomics, work and health* 1991 Macmillan ISBN 0 333 48998 5
Pheasant S and Stubbs D *Lifting and handling - an ergonomic approach* 1991 National Back Pain Association ISBN 0 9507726 4 X

Names of ergonomics practitioners and consultants:
The Ergonomics Society, Devonshire House, Devonshire Square, Loughborough, Leicestershire LE11 3DW.
The Institute of Materials Management, Cranfield Institute of Technology, Cranfield, Bedfordshire MK43 OAL

Examples of industry-specific guidance
HSC *CONIAC condemns heavy building bricks* Construction Industry Advisory Committee, HSC News Release C48:89, 23 November 1989 (revised version under discussion)

European Coal and Steel Community *Guidelines for manual handling in the coal industry* Community Ergonomics Action Luxembourg Report No 14 Series 3 1990

European Coal and Steel Community *Guidelines for manual handling in the steel industry,* Community Ergonomics Action, Luxembourg Report No 16 Series 3 1991

HSC's Health Services Advisory Committee *Guidance on manual handling of loads in the health services* HMSO 1992 ISBN 0 11 886 3541

Other legislation
Management of Health and Safety at Work Regulations 1992, SI 1992 No 2051 HMSO 1992 ISBN 0 11 025051 6
Personal Protective Equipment at Work Regulations 1992 SI 1992 No....(not known at time of going to print)
Provision and Use of Work Equipment Regulations 1992 SI 1992 No....(not known at time of going to print)
Workplace (Health, Safety and Welfare) Regulations 1992 SI 1992 No....(not known at time of going to print)
Control of Substances Hazardous to Health Regulations 1988 SI 1988 No 1657 HMSO 1992 ISBN 0 11 087657 1

Printed in the United Kingdom for HSE, published by HMSO   C200   12/92